Chaparrals

Michael de Medeiros

WEIGL PUBLISHERS INC.

Published by Weigl Publishers Inc.
350 5ᵗʰ Avenue, Suite 3304, PMB 6G
New York, NY 10118-0069
USA

Web site: www.weigl.com

Library of Congress Cataloging-in-Publication Data

de Medeiros, Michael.
    Chaparral / Michael de Medeiros.
    p. cm. — (Biomes)
    Includes index.
    ISBN 1-59036-438-4 (hard cover : alk. paper) —
    ISBN 1-59036-439-2 (soft cover : alk. paper)
    1. Chaparral ecology—Juvenile literature. I. Title. II. Biomes (Weigl Publishers)
QH541.5.C5D46 2006   577.3'—dc22   2006001034

Printed in China
1 2 3 4 5 6 7 8 9 0 10 09 08 07 06

Project Coordinator
Heather Kissock

Designers Warren Clark,
Janine Vangool

Cover description: Yuccas
are a common plant in the
California chaparral.

All of the Internet URLs given
in the book were valid at the
time of publication. However,
due to the dynamic nature of
the Internet, some addresses
may have changed, or sites may
have ceased to exist since
publication. While the author
and publisher regret any
inconvenience this may cause
readers, no responsibility for
any such changes can be
accepted by either the
author or the publisher.

CONTENTS

Introduction

Earth is home to millions of different **organisms**, all of which have specific survival needs. These organisms rely on their environment, or the place where they live, for their survival. All plants and animals have relationships with their environment. They interact with the environment itself, as well as the other plants and animals within the environment. This interaction creates an **ecosystem**.

Different organisms have different needs. Not every animal can survive in extreme climates. Not all plants require the same amount of water. Earth is composed of many types of environments, each of which provides organisms with the living conditions they need to survive. Organisms with similar environmental needs form communities in areas that meet these needs. These areas are called biomes. A biome can have several ecosystems.

The terrain of the chaparral biome is often rugged in appearance.

Chaparral biomes are found in many parts of the world. The word *chaparral* comes from a Spanish word that means scrub oak. The biome is best known for its dense growth of shrubs and its very hot, dry climate. This heat is a major reason that every 40 years or so, huge fires destroy large amounts of chaparral.

Many plants, animals, insects, and other living organisms make their home in the chaparral. This makes it one of the most interesting places on Earth to study. Many scientists study the chaparral biome.

Yucca plants are one of the plants found in the Angeles National Forest in California.

FASCINATING FACTS

Most of coastal and inland California is chaparral.

The chaparral biome is covered with bushes that grow close together. Sometimes the growth is so dense that neither people nor animals can walk through it.

Chaparral is the smallest biome on Earth.

Chaparral Locations

Five of Earth's seven continents have a chaparral biome. Chaparral biomes can be found in North and South America, Australia, Africa, and Europe. They are found in the southwestern coast of the United States, the western and southern parts of Australia, the coast of the Mediterranean, the Cape Town region in South Africa, and the western Chilean coast of South America. Each of the regions is different. Some chaparral biomes are on hilly areas covered in rocks, others are in flat, earthy land areas, and some are in mountainous regions.

The chaparral biome is only found in areas between 30° and 40° latitude. It is almost always on the western side of a continent. On the western side of the continents, there is an ocean that can supply the biome with cooler air and moisture.

Chaparral areas in Europe begin near the seashore and extend inland to mountainous regions.

The Sierra Nevadas form a natural barrier along California's eastern border.

Two chaparral areas that have been studied extensively are the Mediterranean and California chaparral biomes. The Mediterranean chaparral is found on the coast of the Mediterranean Sea, which includes a large area of land reaching parts of North Africa and Europe.

The California chaparral is located on the coast of the southwestern state of California in the United States. It extends into nearby areas, including the Sierra Nevada mountain range. The California chaparral includes mountains and flat lands, making it one of the most interesting areas in the world to study. The fact that the California chaparral is so large and varied means that a great number of different animals, plants, insects, and other living organisms make their home in the area. Unfortunately, much of the California chaparral biome has disappeared because of expanding populations.

FASCINATING FACTS

Chaparral biomes are usually found between desert and forest areas or between desert and grassland areas.

Many movie and television westerns, including early shows starring Roy Rogers, have been filmed in the California chaparral.

WHERE IN THE WORLD?

The chaparral biome is found on almost every continent in the world. Looking at the map, can you spot any of the world's chaparral areas? Find where you live, and see how close you are to a chaparral biome.

Arctic Ocean

North America

Pacific Ocean

Atlantic Ocean

South America

N

0   1000   2000 kilometers

0   500   1000 miles

Chaparral

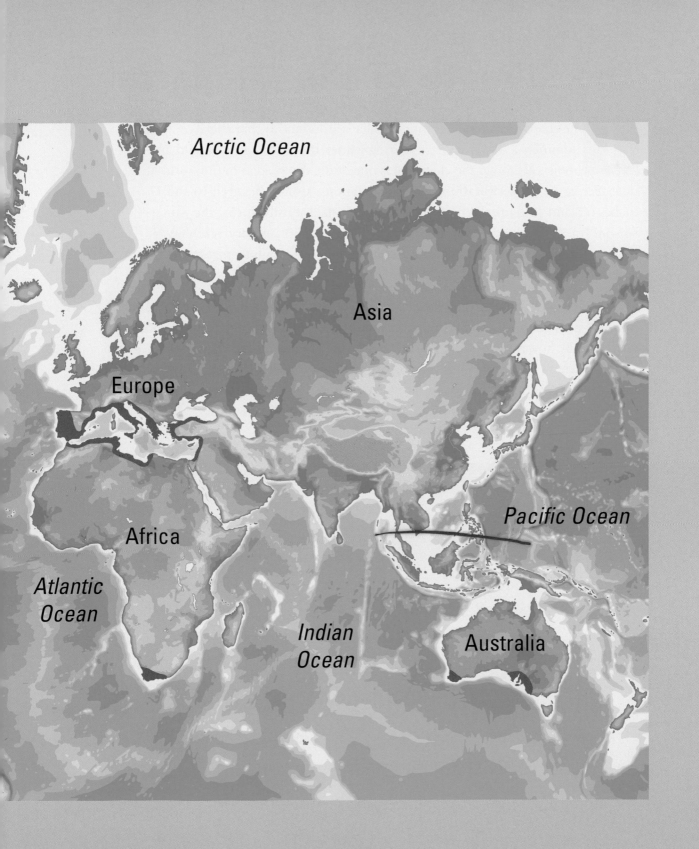

The Climate of the Chaparral

The chaparral biome has four distinct seasons and is characterized by dry summers and wet winters. During winter, the climate is very mild and moist. Although the chaparral is one of the driest areas in the world, it does receive some **precipitation**. This precipitation—usually between 10 and 17 inches (25 and 42 centimeters) for a whole year—falls mostly during the winter season. For the rest of the year, the biome is incredibly hot and dry, reaching temperatures of 100° Fahrenheit (38° Celsius). Due to the high temperatures and dry air, fires and drought are a constant hazard at this time of year.

The Mediterranean chaparral follows this general pattern of high temperatures and minimal precipitation. Much of this area's moisture comes from fog from the oceans nearby. Rain, however, does fall. During winter, the Mediterranean chaparral receives about 7 inches (17.5 cm) of rain. Spring brings about 2 inches (5 cm) of rain. Less than 1 inch (2.5 cm) is received in summer, and about 4 inches (10 cm) of rain falls in the region during the autumn months.

California Chaparral Climatogram

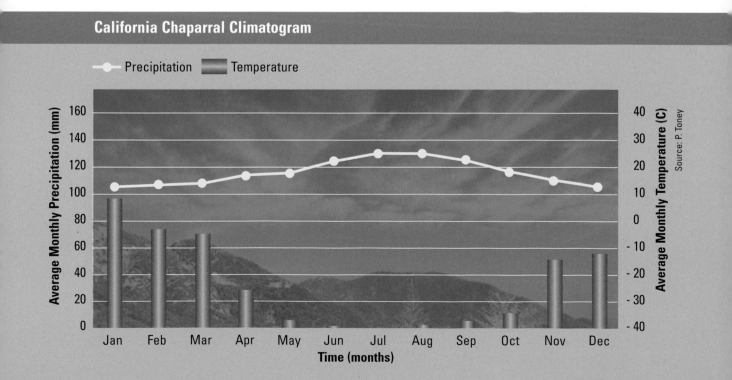

The chaparral on California's coast receives moisture from fog off the Pacific Ocean. The interior is much drier.

The climate in the California chaparral biome is much like the Mediterranean region. However, the higher altitudes of the mountains are cooler and receive more precipitation than the lower regions. In fact, at its highest points, the California chaparral sometimes experiences freezing temperatures. This does not last very long, and any snow that falls melts quickly. Most of the moisture in this area comes in the form of rain, with the region receiving 12 to 40 inches (30 to 102 cm) annually. The rainfall is evenly distributed among the autumn, winter, and spring seasons.

FASCINATING FACTS

The rainy season for the chaparral biome peaks in February. Precipitation slows down after February and completely stops in May.

The climate of the chaparral is often called the Mediterranean climate.

Thunderstorms occur often in the chaparral biome. Fire is a risk, as lightning sometimes strikes.

Chaparral Features

Most chaparral biomes formed as a result of human activity. These areas were once covered with forests. As human populations spread, the land was needed for other purposes. The forests were cut down to make room for farm animals to graze. Over time, this grazing caused the soil to lose many of its **nutrients**. The forests could not grow again because of the damaged soil. The soil became vulnerable to **erosion** by wind and rain. Erosion removed even more nutrients from the area, making it difficult for the soil to become fertile again. Only thick, tough vegetation can grow in these conditions. The chaparral plants overtook the land, and the forests were unable to return. As a result, the chaparral is known for its nutrient-poor soil and shrublike plant life.

The plant life of the chaparral can be divided into two groups. Hard, or true, chaparral is normally found in dry, inland areas that receive little precipitation, including fog. Soft chaparral is usually found in fog belt areas, but can also develop in areas where hard chaparral has experienced fire. In these instances, the soft chaparral is a temporary growth and is replaced by hard chaparral over time.

The king sugarbush is a hard chaparral plant. It resprouts from underground after a fire.

FASCINATING FACTS

During World War II, groups opposing Nazi rule sprang up across Europe. A group of French resistors called themselves the Maquis after the dense bush that they used for cover.

The oils in many soft chaparral plants are heavily scented. This repels animals that might otherwise feed on them.

The roots of some hard chaparral plants extend as far as 30 feet (9 meters) into the ground.

Types of Chaparral

The location of a chaparral biome often determines the type of plant growth that occurs.

Maquis Chaparral

Also known as the chamiso-redshank, the maquis is the most common type of chaparral. Found in both Europe and California, this type of chaparral contains both large and small shrubs as well as a variety of small plants. All of this vegetation is packed tightly together in a dense mass.

Garigue Chaparral

Garigue chaparral is found mainly in Europe. In these areas, the soil is so poor in nutrients that large shrubs cannot grow. Instead, this chaparral is characterized by the growth of herbs and small shrubs. This type of chaparral often moves into areas of maquis chaparral that human development has destroyed.

Coastal Scrub Chaparral

Coastal shrub chaparral is found in the coastal areas of California. It tends to grow in areas that have been destroyed by humans. Coastal scrub chaparral often consists of soft chaparral plants.

Mixed Chaparral

Mixed chaparral is found in California. Located on mountainsides with elevations up to 5,500 feet (1,676 m), mixed chaparral contains shrubs and trees that grow to about 15 feet (5 m).

Montane Chaparral

Montane chaparral is found on the mountains of California. In montane chaparral, large shrubs and dwarf trees grow at elevations up to 9,000 feet (2,743 m). These trees and shrubs rarely grow taller than 10 feet (3 m).

Technology in the Chaparral

The chaparral biome faces unique challenges in survival. Due to its poor soil, only certain types of plants can grow in the area. Yet, even these plants are in danger of disappearing due to air pollution, fire, drought, and human activities. Scientists constantly study chaparral regions to assess the dangers that threaten the biome. Global Positioning System/Global Information System (GPS/GIS) technologies, remote sensing, and **automated** field mapping techniques are just a few of the methods used to track the vegetation of the chaparral.

GPS and GIS are both used to monitor plant growth in the chaparral. GPS locates where the plants are growing in the biome. GIS tracks which plants are growing where. By knowing where certain plants are growing, scientists can analyze the area to determine the conditions needed to grow these plants. It can also help them understand why other plants do not grow in the area. GPS also alerts scientists to the disappearance of some plants from an area. They can then study the factors that led to the disappearance and suggest solutions to improve the environment so the plants will return.

Remote sensing also helps scientists track plant growth in the chaparral. Like GPS, remote sensing uses satellites to locate plants. In remote sensing, the satellites provide **infrared** images that show scientists the plants growing in an area, as well as all of the other objects found in the area as well. This, again, helps scientists assess the conditions under which certain plants grow. It also helps them see how changes to the environment, including pollution levels and temperature fluctuations, affect plant growth.

Satellites monitor weather conditions in California, including the Sierra Nevadas, to warn of potential fires.

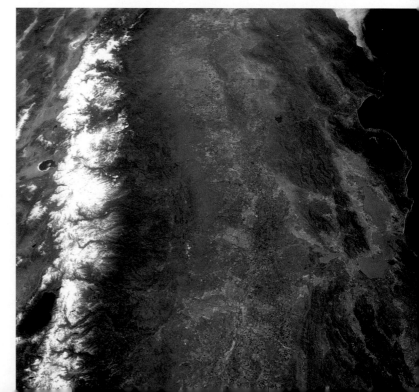

To properly understand the information obtained from GPS/GIS and remote sensing, scientists must have a place to put it when it is gathered. Computer-automated field mapping programs have been developed for this purpose. They allow scientists to create a picture of a chaparral biome. Scientists can see which plants grow in the biome, where these plants are growing, and what other plants, animals, and objects share the biome with them. They can also map the conditions affecting the biome, including air quality and human development. By putting all of the information together, the scientists can properly assess and address the issues facing the biome.

Remote sensing and satellite imagery are also used to track earthquake activity.

FASCINATING FACTS

GIS has a variety of uses, including emergency planning. By tracking emergency calls and travel times, the system can be used to calculate response times during disasters and other crises, including chaparral fires.

Remote sensing is often used to track the effects of flooding by comparing normal water levels to rising water levels.

LIVING THINGS

The chaparral biome is the natural habitat of many different plants and animals. The plants and animals that survive in these areas have learned to adapt to the conditions. They know how to live on very little water and how to conserve water as much as possible.

The roadrunner can run as fast as 17 miles (27 kilometers) per hour.

BIRDS

More than 100 different kinds of birds live in the chaparral region of California. Living in this hot region has forced these birds to adapt to the heat. Some birds that live in this biome are the rock dove, house sparrow, yellow-billed cuckoo, rough-winged swallow, black-chinned hummingbird, roadrunner, loggerhead shrike, cactus wren, and the California quail.

MAMMALS

Like birds, some animals have adapted to the harsh environment of chaparral biomes. Two animals commonly found in the California chaparral are the gray fox and the spotted skunk. The golden jackal and Bezoar goat can be found roaming the Mediterranean chaparral.

Golden jackals inhabit open country with trees, brush, or grass.

Honeybees are social insects. Their nests, or hives, contain up to 20,000 bees.

INSECTS

Many different kinds of insects can be found in the chaparral biome. Flies, honeybees, and ladybugs flit about the chaparral biomes. Cockroaches make their home under rocks. Butterflies and moths **pollinate** the hardy flowers of the chaparral.

PLANTS

Vegetation in the chaparral biome is mainly composed of different shrub species. These plants have adapted to living with the fires, heat, and dryness in the region. Two common shrubs found in the biome are the manzanita and the chamise. The chaparral biome is also home to many **herbs,** such as oregano, rosemary, and thyme.

Oregano has a strong aroma and a slightly bitter flavor.

Chaparral Plants

Shrubs

The landscape of the chaparral is characterized by its dense shrub growth. Two of the most common shrubs found in the California chaparral are the chamise and the greenleaf manzanita. The chamise is an evergreen shrub that grows to between 3 and 10 feet (1 and 3 m) in height. When caught in a brush fire, it burns quickly and intensely, much like grease. For this reason, it is sometimes known as "greasewood." Chamise can be used to make tea and was used by American Indians and settlers to treat diseases, such as tetanus and rabies. Greenleaf manzanita is often found in montane chaparral. The word *manzanita* means "little apple" in Spanish. The plant was given this name because of its small red and green fruit. Scotch broom is a shrub common to the European chaparral. Known for its rapid spread, the plants make a popping sound as seeds burst from their pods.

Beavertail cacti bloom from March to June.

Cactus

Beavertail cactus and Lee pincushion cactus are two species of cactus that grow in the California chaparral. The stems of the beavertail are flat and paddlelike, resembling the tail of a beaver. They are blue-green in color for most of the year, but sometimes turn deep purple with the approach of winter. Beavertail cactus blooms in the spring with flowers that are deep rose. Lee pincushion cactus grows in interior chaparral regions. It is normally found growing in small cracks in the ground.

A greenleaf manzanita can grow up to 5 feet (1.5 m) in height.

Sage is commonly thought to improve memory.

Herbs

Oregano, sage, rosemary, and thyme are a few of the herbs that grow in the chaparral regions of the world. Oregano is an herb that originated in the Mediterranean. It is a common ingredient in Greek and Italian cooking. A member of the mint family of plants, the word *oregano* means "joy of the mountains." Sage plants grow to about 3 feet (1 m) in height. This herb is commonly used to season turkey and other meats. As a tea, it can also be used to fight a cold. Rosemary grows wild in the Mediterranean region. Like oregano and sage, this herb is used in cooking. Rosemary is thought to have medicinal uses. Some people use it to relieve aching joints. Thyme is another herb native to the Mediterranean region. It has a lemon flavor and is often used in French cooking. Like sage, it is used to ward off a cold.

FASCINATING FACTS

In the Middle Ages, sage was often added to meat that was going bad. It masked the rotten taste and helped in the digestion of the food.

As a flower, rosemary symbolizes remembrance. In ancient Greece, students put the plant in their hair when studying for exams.

In the past, American Indians found that the chamise was useful when making arrows. The Koso people used it for arrow points, while the Luiseno people used it for the arrow shaft.

Poison oak is another shrub found in the California chaparral. A single touch to its leaves can cause a severe rash.

Chaparral Birds and Mammals

The Cactus Wren

One of the most common birds in the California chaparral is the cactus wren. It is a small bird—only about 9 inches (22 cm) long. However, it is the largest wren in the United States. This curious bird is as active as it is eye-catching. It may not be very colorful, but the white on its throat, belly, and sides is a startling contrast to its spotted breast.

The cactus wren is Arizona's state bird. The bird can be frequently spotted on top of a saguaro cactus.

The California quail is sometimes called the valley quail.

The California Quail

The California quail is California's state bird. It is a favorite of birdwatchers and hunters. The quail makes its nest in brushy or grassy places. The quail is a **monogamous** bird. It has only one mate over the course of a breeding season.

The Bezoar Goat

The Bezoar goat is commonly referred to as the wild goat. It can be found on many Greek islands, as well as in Turkey, Pakistan, and Iran. The Bezoar goat weighs up to 300 pounds (136 kilograms), but this weight is often hidden by the thick wool that covers its body. The most common colors for the Bezoar goat are white, gray, brown, red, and black. The horns of Bezoar goats grow in the shape of a **scimitar**.

The Gray Fox

The gray fox is commonly known as the tree fox. It lives in the southwestern United States and can be found in the California chaparral. The gray fox prefers to live in wooded areas that are covered with bushes. It is a nocturnal animal, so it sleeps during the day and hunts at night. Being an **omnivore**, the gray fox eats nuts and berries and also hunts small animals, such as rats and rabbits. A gray fox can live for about 12 years in captivity, but its lifespan in the hot, dry chaparral is much shorter.

The gray fox is the only member of the dog family that can climb trees.

FASCINATING FACTS

The California chaparral biome is the habitat of many kinds of hawks, including red-tailed hawks and sharp-shinned hawks. These birds normally leave the chaparral when the hot weather arrives.

The gray fox can reach speeds of 28 miles (45 km) per hour.

Wild female and baby goats live together in packs of almost 50. Male goats either live by themselves or in an all-male pack.

Insects of the Chaparral

Butterflies and Moths

A butterfly common to the California chaparral is the pale swallowtail. This butterfly is easily recognized. Its cream-colored base is marked with heavy black stripes and borders. The pale swallowtail is mostly found in hilly areas, where it feeds on the nectar of the chaparral's many plants. The sphinx moth is also called the hawk or hummingbird moth because of the way it hovers in the air. Its wingspan can be more than 5 inches (12.5 cm), making it one of the largest flying insects in the chaparral region. The yucca moth is the only insect that pollinates yucca plants. It does this by stuffing a small ball of **pollen** into the center of each flower.

Stink beetles are also known as "clown head" beetles because of the way they respond to danger.

Cockroaches and Beetles

The western wood cockroach resides in or under the logs and rocks of the chaparral. These cockroaches are small in size. Males range from 0.47 to 0.55 inches (1.2 to 1.4 cm) in length, and females range from 0.35 to 0.51 inches (0.9 to 1.3 cm). While the females are wingless, the males have wings and are able to fly. Both feed on decaying plant matter found in the area. The yucca weevil is a beetle that feeds off the yucca plant. The larvae of the weevil tunnel under the bark of the plants. This can sometimes destroy the plant. Stink beetles have a unique way of responding to danger. They perform a handstand and point their abdomen at the threat. They then run away in this position.

A female yucca moth pollinates the yucca plant by injecting three to five eggs into the plant.

The average life span for a
Mediterranean fruit fly is 30 days.

Flies and Midges

The buzz of flies and midges is constant
on the chaparral. Scorpionflies are often
mistaken for giant mosquitoes. The
main difference between the two,
however, is the scorpion-like "tail" that
hangs from the abdomen of some males.
It is this tail that gives the scorpionfly its
name. As their name suggests, fruit flies
feed on the **nectar**, pollen, and seeds of
fruit. Some fruit fly species spend their
entire lives living on a single plant.
Others travel hundreds of miles in their
lifetime. Gall midges look like small
mosquitoes. These tiny flies feed on
plants. This feeding causes the plant to
swell, forming bumps called galls. The
galls rarely harm the plant.

FASCINATING FACTS

When a stink beetle is picked up,
it releases a foul-smelling, reddish
black liquid. This is how the beetle
received its name.

California carpenter bees, another
chaparral resident, bore holes into
trees. They raise their young in
these holes.

Danger in the Chaparral

The chaparral biome is under constant threat of fire. Every 30 to 40 years, the chaparral experiences natural fires that destroy large areas of land and the nutrients within. Fire is a necessity to this biome. It rids the area of dense plant growth that has been on the land for years. This allows new growth to emerge. Fires allow the chaparral to regenerate.

Natural fires, however, can quickly become uncontrollable. Chaparral plants are dry and burn quickly. The fire can quickly spread a great distance. Fires in the chaparral have been known to move at a rate of 8 miles (13 km) per hour. These fires quickly consume everything in their path.

While fire is good for the chaparral in many ways, it also has an adverse effect on the biome. Plants that have been growing in the chaparral for a while become rooted firmly in the soil. These plants and roots help hold the soil in place. When fire destroys the dense plant growth, the barrier that was protecting the soil is removed. The soil and nutrients can be easily shifted and washed away. This makes the ground water resistant. Since the water docs not soak in, it remains very dry.

For people living in or near the chaparral, fires are a major concern. Both professional and volunteer firefighters are called upon to help control the fire's spread.

A yucca plant blooms among the charred branches in Lytle Canyon, California, after a major brush fire burned its way through the area.

To keep the problem under control, some governments are letting people start fires in small parcels of land. This process is called controlled burning. It reduces the risk of the fire spreading into and overtaking farmland and other nearby residential communities. The California Department of Forestry has allowed ranchers to start controlled fires in the California chaparral for almost 50 years. Still, as much as controlled burning helps stop massive fires, they do still occur. In fact, it is very hard to predict when and where a fire will break out in the chaparral biome.

FASCINATING FACTS

The shortage of rain during the summer season is the main reason that fires break out in the chaparral biome.

The plants commonly found in the chaparral hold highly flammable oils, which cause fire.

CHAPARRAL STUDIES

The chaparral biome is one of the most heavily studied biomes in the world. The fires, nutrients, climate, and everyday happenings in the chaparral biome are of great importance to ecologists, climatologists, and soil scientists.

ECOLOGIST

- Duties: studies the relationship between organisms and their environment

- Education: bachelor's, master's, or doctoral degree in science

- Interests: biology, statistics

A love of nature is essential for an ecologist. Ecologists spend much of their time outdoors studying what makes the natural world tick.

They study the relationships between plants and animals to understand how ecosystems work. Ecologists study the life cycles of plants and animals, keeping track of the number and types of plants and animals that live in a certain area. They advise organizations about conservation of ecosystems and perform conservation work themselves.

CLIMATOLOGIST

- Duties: analyzes and forecasts weather, and conducts research into processes and phenomena of weather, climate, and atmosphere

- Education: bachelor's, master's, or doctoral degree in meteorology

- Interests: physics, statistics, ecology, meteorology

Climatologists study climate in any area, including the chaparral regions of the world. By studying the patterns of weather and the changing temperatures of a region, climatologists can understand any region or biome in the world. Their study into weather is so specific that they can even estimate what temperatures and weather in general will be like years from now.

SOIL SCIENTIST

- Duties: studies soils and the implications of soil use

- Education: bachelor's, master's, or doctoral degree in science, Earth science, or a related discipline

- Interests: the environment, microbiology, math, chemistry, geology

Soil scientists investigate soil conditions to determine the biological, physical, and chemical activity taking place in the soil. They can then provide advice on how to restore damaged land and how to maintain these conditions. Soil scientists use their keen problem-solving skills to analyze information and interpret scientific results.

ECO CHALLENGE

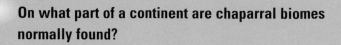

1 On what part of a continent are chaparral biomes normally found?

2 What latitude coordinates are the major locations of chaparral biomes?

3 How hot does the chaparral biome usually get in the summer?

4 How often do major fires sweep through chaparral biomes?

5 At what rate of speed does fire travel in the chaparral biome?

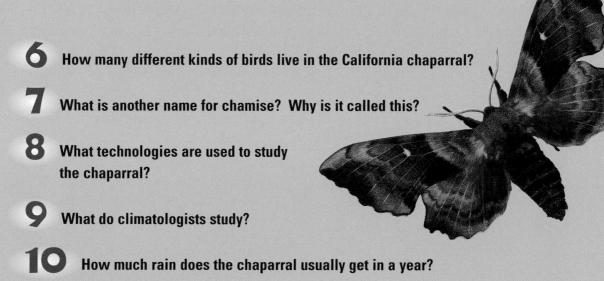

6 How many different kinds of birds live in the California chaparral?

7 What is another name for chamise? Why is it called this?

8 What technologies are used to study the chaparral?

9 What do climatologists study?

10 How much rain does the chaparral usually get in a year?

Answers

1. on the western part of continents
2. between 30° and 40° latitude
3. up to 100°F (38°C)
4. every 30 to 40 years
5. at a rate of 8 miles (13 km) per hour
6. more than 100 different kinds of birds
7. greasewood; because it burns quickly and intensely, like grease
8. GPS/GIS, remote sensing, automated mapping
9. Climatologists study weather and climate.
10. between 10 and 17 inches (25 and 42 cm)

WATERING SOIL

Soil in the chaparral biome is very dry and holds few nutrients. Only certain types of plants can survive in these conditions. The composition of soil affects its ability to hold moisture. When a plant is unable to hold moisture, its ability to support plant life is affected. Try this activity to see how water passes through different types of soil.

MATERIALS

- 3 soil samples (Include a variety of samples, such as garden soil, clay, and sand.)
- 3 jars
- pen
- 3 coffee filters
- 3 funnels
- paper
- timer or stopwatch
- water

1. Put each soil sample in a jar, and label each jar.

2. Place a coffee filter inside each funnel, and place the funnels into the jars.

3. Pour 1/2 cup (125 mL) of each soil sample onto a piece of paper. Weigh each sample. On another piece of paper, create a chart to note the weights.

4. Pour the soil samples into the separate funnels. Set the timer for 30 minutes. Pour 1/2 cup (125 mL) of water into each funnel.

5. Watch to see when the water starts to drip through the filter in the funnel. On your chart, record how long it took for the water to begin dripping from each sample.

6. After 30 minutes, remove the funnels and measure the volume of water that has drained from the sample into the jar. Note the amounts on your chart.

7. Empty each soil sample onto a fresh piece of paper, and weigh it again. Record the weights on your chart. Which samples absorbed and held water the best? Why do you think this happened?

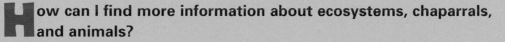

FURTHER RESEARCH

How can I find more information about ecosystems, chaparrals, and animals?

• Libraries have many interesting books about ecosystems, chaparrals, animals, and plants.

• Science centers and research facilities are great places to learn about ecosystems, chaparrals, animals, and plants.

• The Internet offers some great websites dedicated to ecosystems, chaparrals, animals, and plants.

BOOKS

Burnie, David. *Shrublands.* Chicago, IL: Raintree, 2003.

Chaparral. New York, NY: Benchmark Books, 1996.

Halsey, Richard W. *Fire, Chaparral, and Survival in Southern California.* El Cajon, CA: Sunbelt Publications, 2004.

WEBSITES

Where can I learn more about chaparral biomes of the world?

Blue Planet Biomes
www.blueplanetbiomes.org/

Where can I learn more about the animals of the chaparral?

World Wildlife Fund
www.worldwildlife.org

Where can I learn about the many plants of the chaparral?

California Native Plant Society
www.cnps.org/program/
education/chaparral.htm

GLOSSARY

automated: a machine that does a job that used to be done by people

ecosystem: a community of living things sharing an environment

erosion: the process of wearing away by wind, rain, and glaciers

herbs: plants whose leaves, stems, seeds, or roots are used in cooking or medicines

infrared: invisible wavelengths

monogamous: to have only one mate

nectar: sweet liquid formed in flowers

nutrients: substances that provide nourishment

omnivore: an animal that eats plants and other animals

organisms: living things

pollen: a yellow powder made in flowers

pollinate: transfer pollen from one plant to another to fertilize it

precipitation: water in the form of rain, sleet, hail, or snow

scimitar: a curved and deadly sword

vegetation: plant life

INDEX